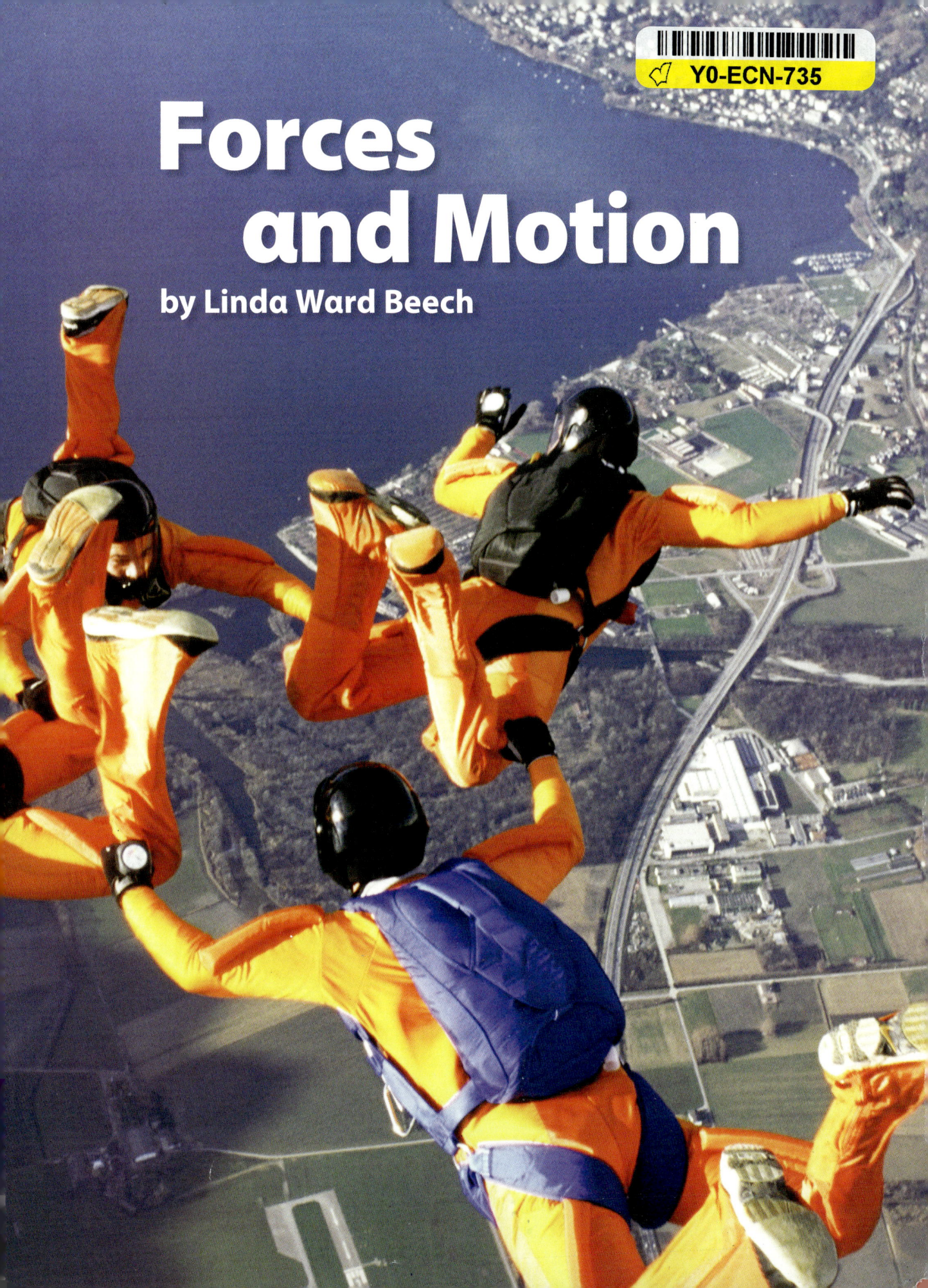

Forces and Motion

by Linda Ward Beech

This ice climber uses force to lower himself.

Contents

Introduction: Moving Things 4

Chapter 1

Big Idea Question
What Is a Force? 6
   Pushes and Pulls 8
   Forces and Motion 10

Chapter 2

Big Idea Question
What Is Gravity? 16
   Gravity 18

Chapter 3

Big Idea Question
What Are Magnets? 22
   Magnets 24

Conclusion: Forces All around Us 28

Glossary 30
Index 32

Introduction

Moving Things

Roller coaster

Next Generation Sunshine State Standards
SC.2.P.13.1 Investigate the effect of applying various pushes and pulls on different objects.

The world is full of moving things. Things move from place to place. An eagle flaps its wings. A roller coaster travels along a track. Musicians beat drums.

Musicians

Chapter 1

Big Idea Question

What Is a Force?

Next Generation Sunshine State Standards
SC.2.P.13.1 Investigate the effect of applying various pushes and pulls on different objects.
SC.2.P.13.4 Demonstrate that the greater the force (push or pull) applied to an object, the greater the change in motion of the object.

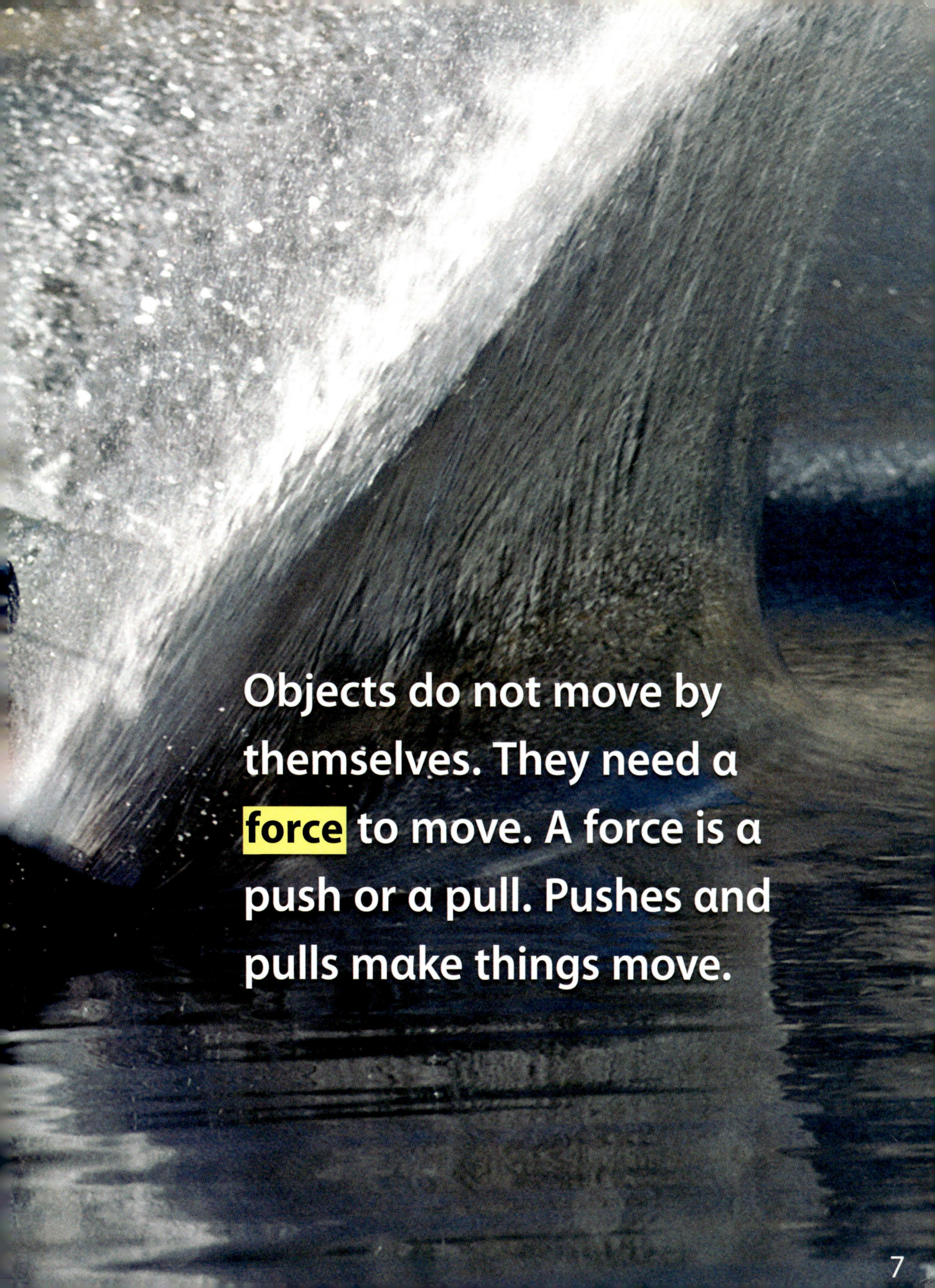

Objects do not move by themselves. They need a **force** to move. A force is a push or a pull. Pushes and pulls make things move.

Pushes and Pulls

A push moves something away from something else. These men push a boat into the water. The push moves the boat away from the ice.

A pull moves something toward something else. This boy pulls a post. It moves toward him. The pull changes the position of the post.

Forces and Motion

When something moves, it is in **motion**. Forces put objects in motion.

When a bowling ball knocks pins over, the ball and the pins move. They are in motion.

Motion is not always the same. When you kick a ball, you use force to move it. If you kick the ball lightly, it will not go very far.

Less Force

If you kick the ball harder, it can go farther. The ball can also go faster. That's because you are pushing harder.

More Force

This boy uses force to move a truck. The toy truck is empty and light. He gives it a push to move it forward.

The boy fills the truck with sand. Now it is heavier. He uses the same amount of force to push the truck. Which one could go farther?

Chapter 2

Big Idea Question

What Is Gravity?

Next Generation Sunshine State Standards
SC.2.P.13.3 Recognize that objects are pulled toward the ground unless something holds them up.

Have you ever seen a waterfall? Water rushes over cliffs and crashes to the ground. What makes this happen? Earth's **gravity** moves the water and makes it fall.

Gravity

Gravity is a pulling force. Earth's gravity pulls things toward the center of the Earth. Gravity pulled these apples down to the ground without touching them.

Gravity pulls objects down unless something holds them up. This girl lifts a basket of apples. Her lift keeps the basket of apples from falling to the ground.

Gravity pulls on objects through air. These leaves fall to the ground. Gravity pulls them down through the air. Gravity also pulls on objects through liquid. This anchor is pulled down through the water to the ocean's floor.

Gravity pulls on objects through solids. A solid can keep an object from falling to the ground. Gravity pulls this plant toward the ground, but the solid table holds it up.

Chapter 3

Big Idea Question

What Are Magnets?

Magnets push and pull. They only pull on certain metal objects, like these iron filings.

Magnets

A magnet has two **poles**. A magnet has a north pole and a south pole. A magnet's force is strongest at its poles.

The N stands for the north pole.
The S stands for the south pole.

A north pole and a south pole are opposite poles. They **attract,** or pull toward each other. Two north poles or two south poles are alike. They **repel,** or push away from each other.

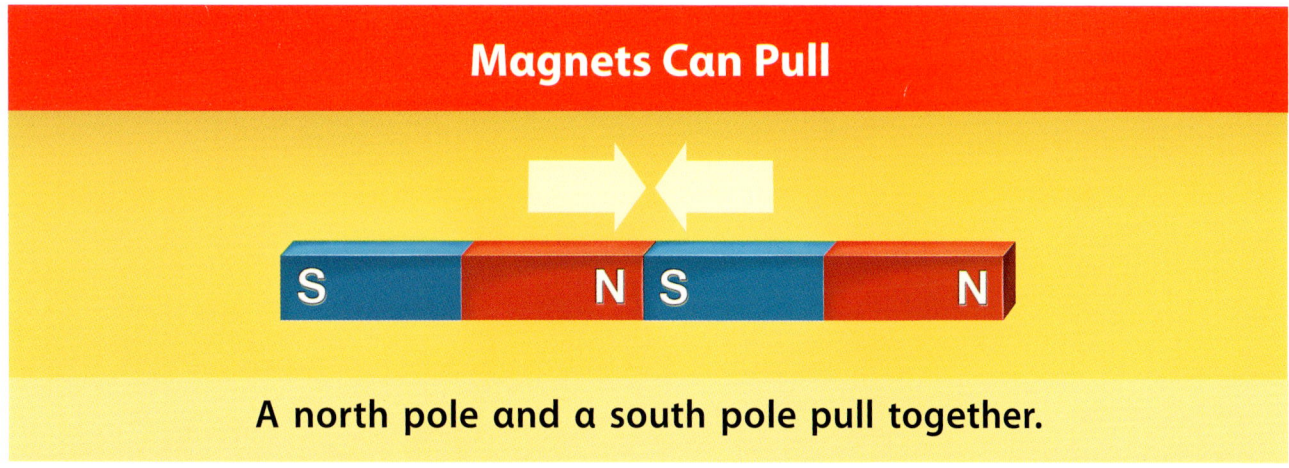

A north pole and a south pole pull together.

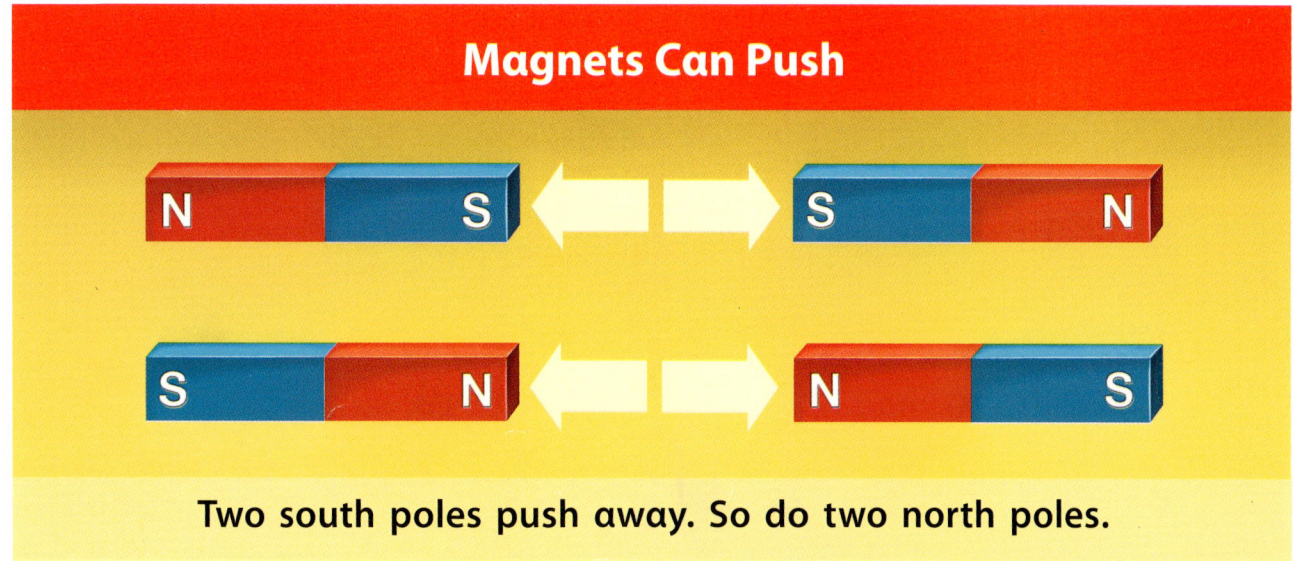

Two south poles push away. So do two north poles.

Magnets can pull objects through air, liquids, and solids. A magnet pulls these iron nails through the air. Magnets can also attract nails through liquid.

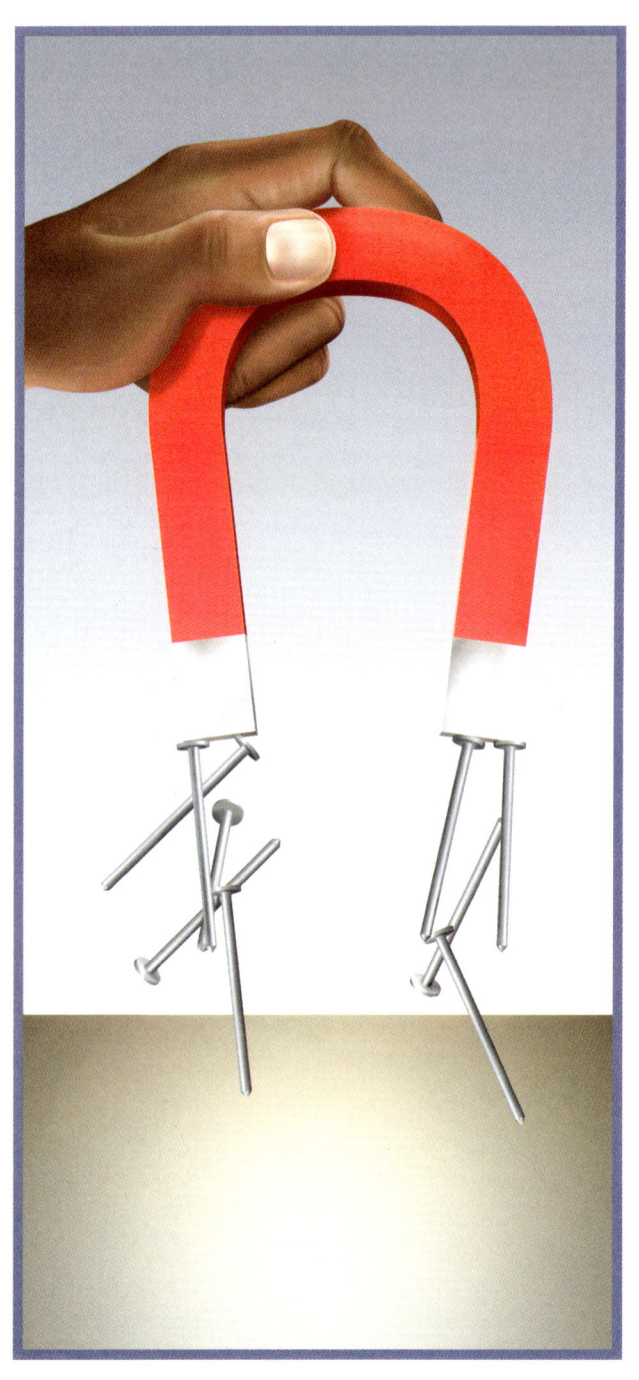 

A magnet can even pull things without touching them. This magnet attracts nails through a solid board.

Conclusion

Forces All around Us

Objects need a force to move them. They need a push or a pull. Forces make things move.

Earth's gravity is a force that pulls everything down toward the center of Earth. Objects will fall toward the ground unless something holds them up.

Magnets pull objects made of some metals. Magnets can make things move without touching them.

Next Generation Sunshine State Standards
SC.2.P.13.1 Investigate the effect of applying various pushes and pulls on different objects.
SC.2.P.13.2 Demonstrate that magnets can be used to make some things move without touching them.
SC.2.P.13.3 Recognize that objects are pulled toward the ground unless something holds them up.

Glossary

attract (page 25)
To **attract** is to pull toward.

A north pole and a south pole **attract,** or pull together.

force (page 7)
A **force** is a push or a pull.

The water skier uses **force** to move through the water.

gravity (page 17)
Earth's **gravity** is a force that pulls things to the center of Earth.

Earth's **gravity** moves the water and pulls it to the ground.

magnet (page 23)
A **magnet** is an object able to pull some metals toward itself.

The **magnet** pulls iron filings.

motion (page 10)

When an object is moving, it is in **motion.**

The bowling ball and pins are moving. They are in **motion.**

pole (page 24)

A **pole** is the part of a magnet where its force is the strongest.

The magnet has a north **pole** and a south **pole.**

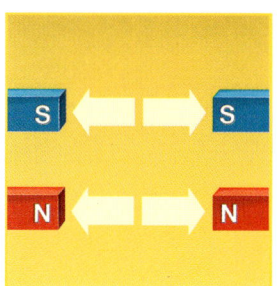

repel (page 25)

To **repel** is to push away.

Two south poles **repel,** or push away, from each other. So do two north poles.

Index

attract 25–27, 30

force 7, 10, 12–15, 18, 24, 28, 30

gravity 16–21, 28, 30

iron 23, 26

magnet 22–27, 30

motion 10–11, 31

pole 24–25, 31

repel 25, 31

Copyright © 2011 The Hampton-Brown Company, Inc., a wholly owned subsidiary of the National Geographic Society, publishing under the imprints National Geographic School Publishing and Hampton-Brown.

All rights reserved. No part of this book may be reproduced or transmitted in any form or by any means, electronic or mechanical, including photocopying, recording, or by an information storage and retrieval system, without permission in writing from the Publisher.

National Geographic and the Yellow Border are registered trademarks of the National Geographic Society.

National Geographic School Publishing
Hampton-Brown
www.NGSP.com

Printed in the USA.
RR Donnelley, Jefferson City, MO

ISBN: 978-0-7362-7560-6

11 12 13 14 15 16 17
10 9 8 7 6 5 4